Inch by Inch

AF710870

Use a ruler. Find the length of the item to the nearest inch.

1. Path M
2. Path N
3. Path O
4. Path P
5. Path Q
6. Path R

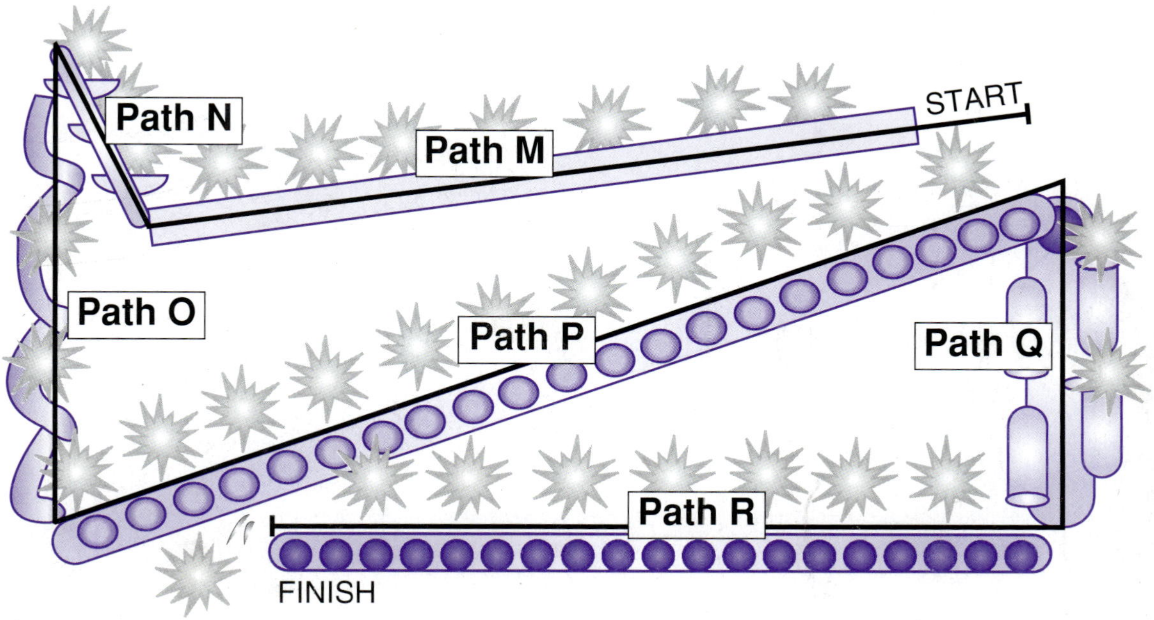

Find the item that is closest to the given length.

7. 2 in.
8. 3 in.
9. 1 in.
10. 5 in.
11. 4 in.
12. 6 in.

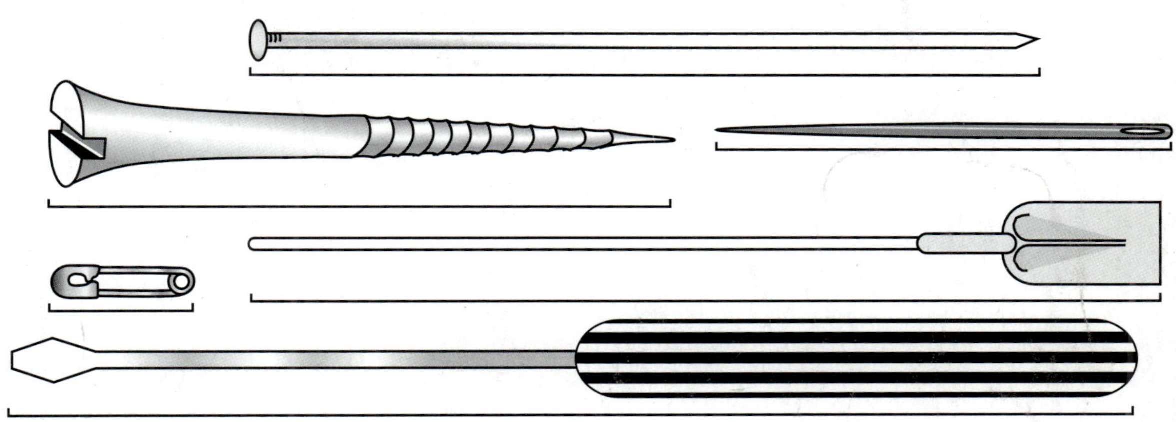

Answer Box

A	B	C	D	E	F
Safety pin	2 in.	1 in.	4 in.	Shovel	6 in.
G	**H**	**I**	**J**	**K**	**L**
Nail	3 in.	Screw driver	Screw	5 in.	Needle

Objective: Measure an object to the nearest inch.

Nature Rules!

Use a ruler. Find the length of the item to the nearest inch.

1. 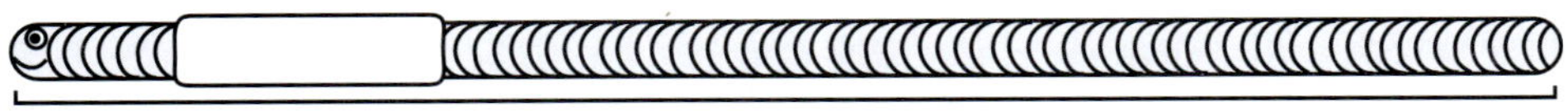

2. 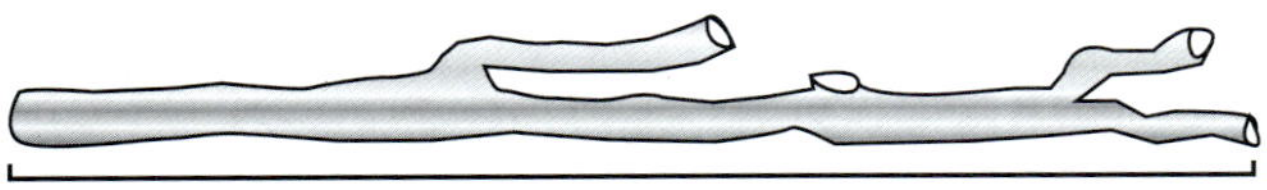

3. 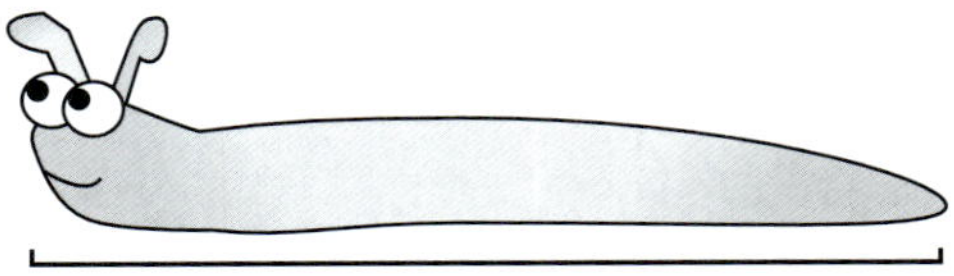

4. 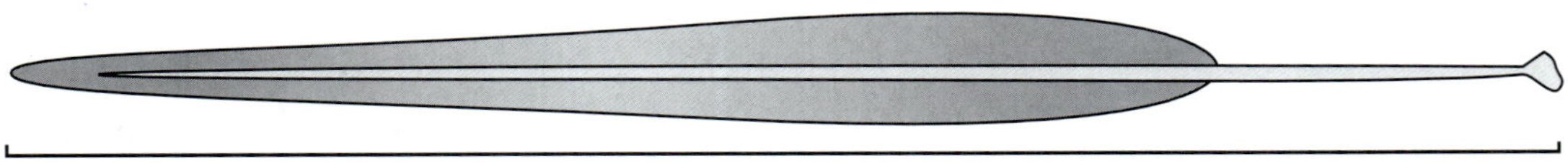

5. 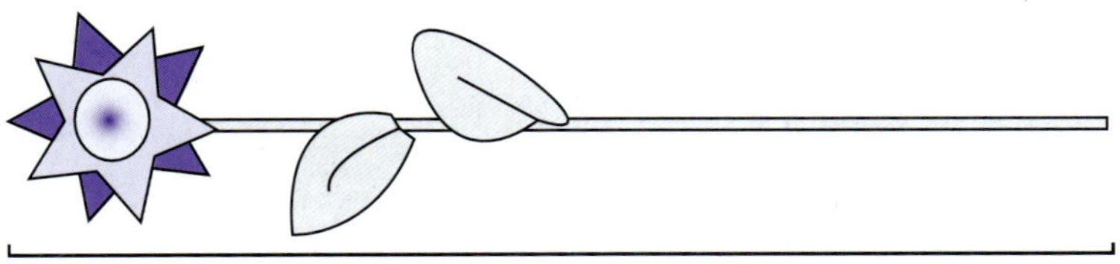

6. 

Find the trail that is closest to the given length.

7. 5 in.
8. 1 in.
9. 3 in.
10. 6 in.
11. 4 in.
12. 2 in.

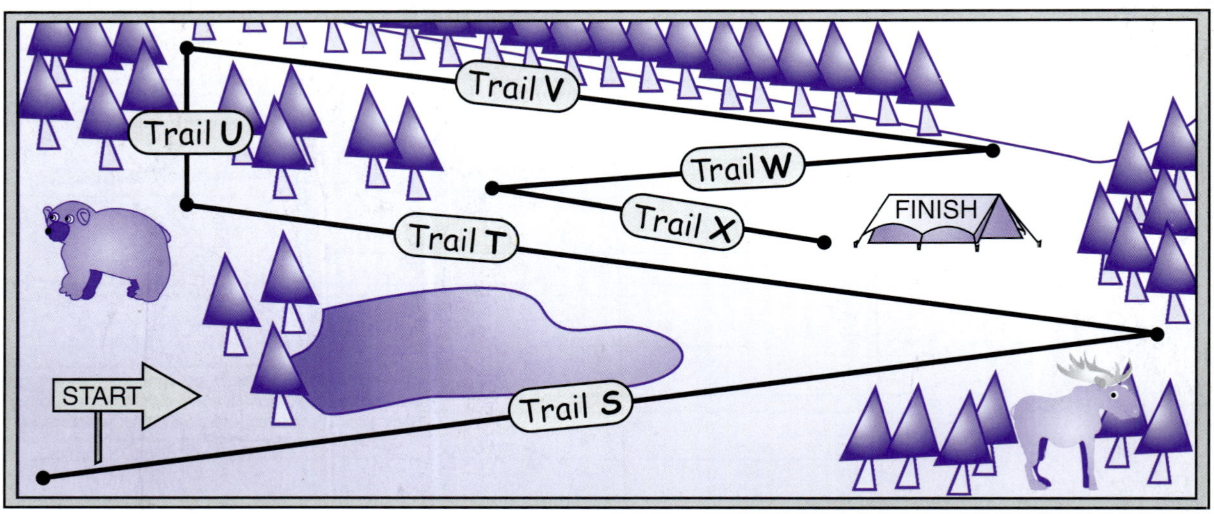

Answer Box

A	B	C	D	E	F
Trail X	Trail U	4 in.	Trail V	Trail S	Trail T
G	**H**	**I**	**J**	**K**	**L**
3 in.	1 in.	2 in.	Trail W	6 in.	5 in.

 Objective: Measure an object to the nearest inch.

That's Close!

Example

Choose the best estimate for the length.

a door
7 in. or 7 ft

Think: 7 inches (in.) is too short.

So, **7 ft** is the best estimate.

Choose the best estimate for the length.

1. a tree
   2 in. or 20 ft
2. a chair
   40 in. or 40 ft
3. a book
   8 in. or 8 ft
4. a crayon
   40 in. or 4 in.
5. a bush
   60 in. or 6 in.
6. a pen
   6 in. or 60 ft
7. a swimming pool
   40 in. or 40 ft
8. a desk
   2 in. or 2 ft
9. an office building
   8 ft or 80 ft
10. a house
    60 ft or 6 in.
11. a ladder
    8 in. or 8 ft
12. an eraser
    2 in. or 20 ft

Answer Box

A	B	C	D	E	F
2 in.	40 in.	40 ft	60 ft	8 in.	8 ft
G	**H**	**I**	**J**	**K**	**L**
6 in.	2 ft	20 ft	4 in.	60 in.	80 ft

Objective: Name the best estimate for the length of an object.

The Long and Short of It!

Choose the best estimate for the length.

1. a sofa
   7 ft or 70 ft

2. a table
   30 ft or 30 in.

3. a driveway
   7 in. or 70 ft

4. a classroom ceiling
   90 ft or 9 ft

5. a pencil
   7 ft or 7 in.

6. a bookcase
   5 ft or 5 in.

7. a postcard
   5 in. or 5 ft

8. a barn
   50 ft or 5 ft

9. a piece of chalk
   3 in. or 30 in.

10. a fence
    9 ft or 90 ft

11. a stapler
    9 ft or 9 in.

12. a school building
    30 ft or 30 in.

Answer Box

A	B	C	D	E	F
90 ft	9 ft	70 ft	3 in.	30 in.	30 ft
G	**H**	**I**	**J**	**K**	**L**
7 ft	5 in.	7 in.	9 in.	50 ft	5 ft

 Objective: Name the best estimate for the length of an object.

Sense With Centimeters

Use a ruler. Find the length of the item to the nearest centimeter.

1 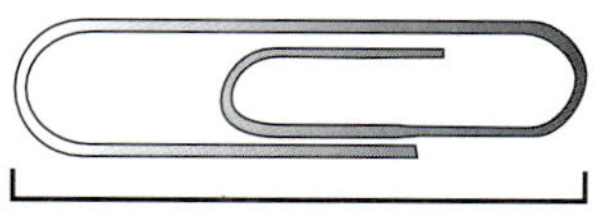

2 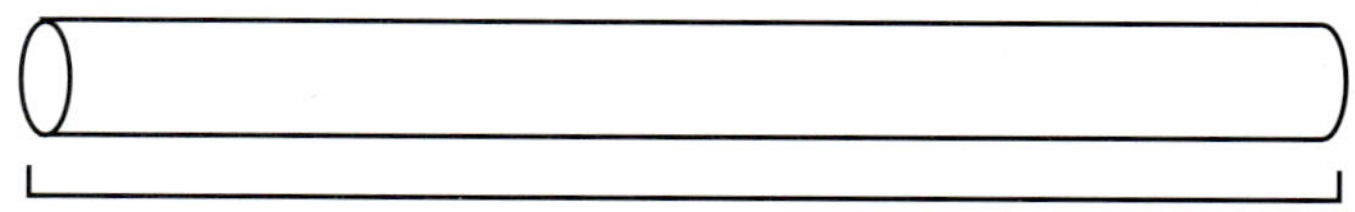

3 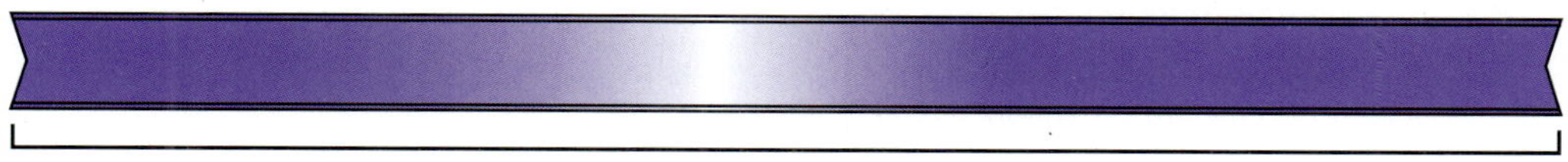

4 

5 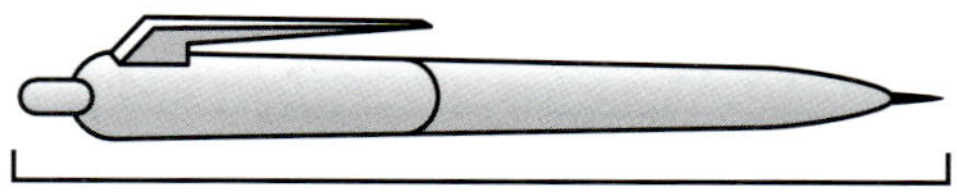

6 

7 

8 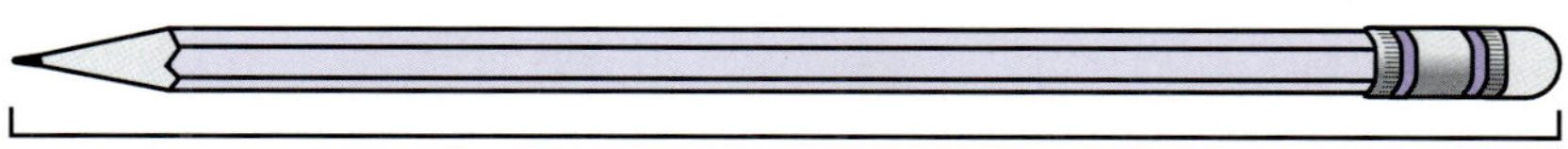

9 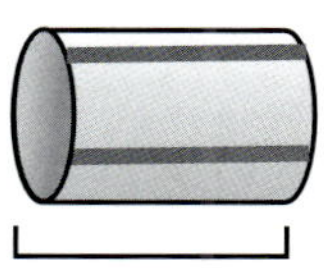

10 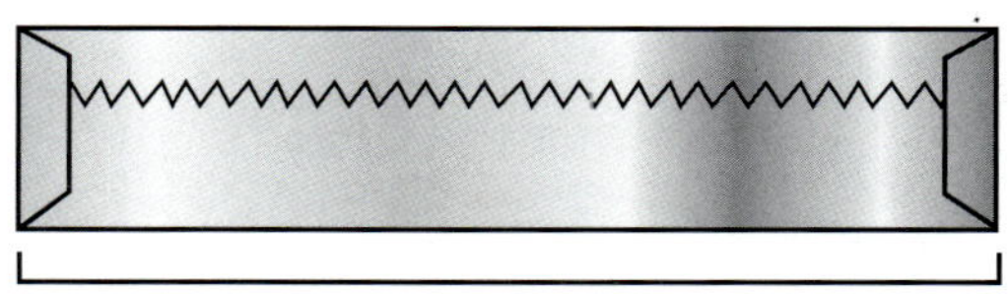

11 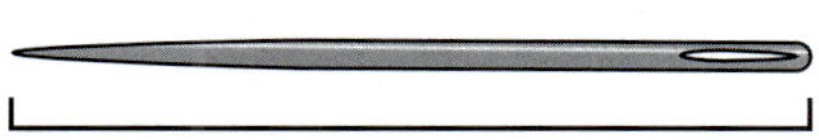

12 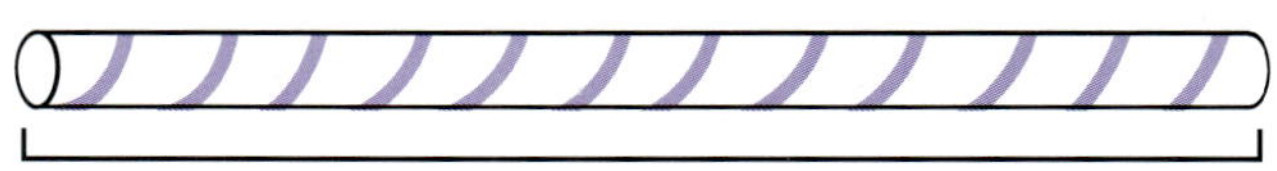

Answer Box

A	B	C	D	E	F
10 cm	4 cm	8 cm	9 cm	6 cm	2 cm
G	**H**	**I**	**J**	**K**	**L**
3 cm	7 cm	12 cm	11 cm	13 cm	5 cm

Natural Lengths

Use a ruler. Find the length of the item to the nearest centimeter.

1. 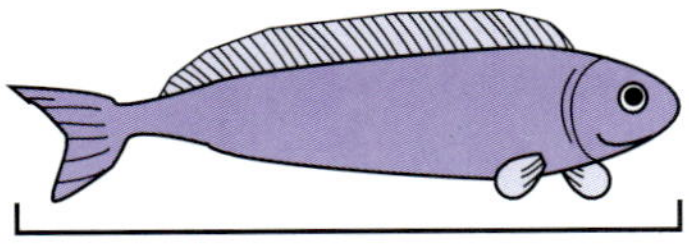
2. 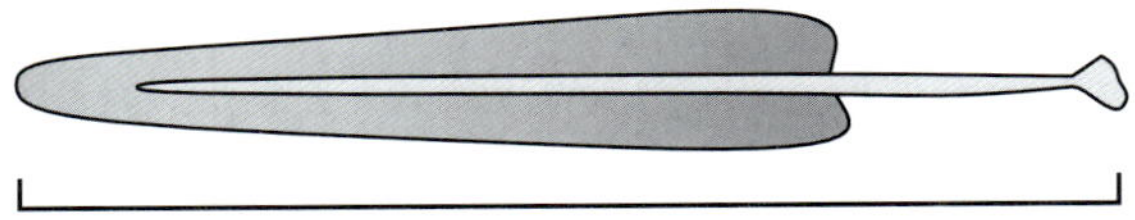
3. 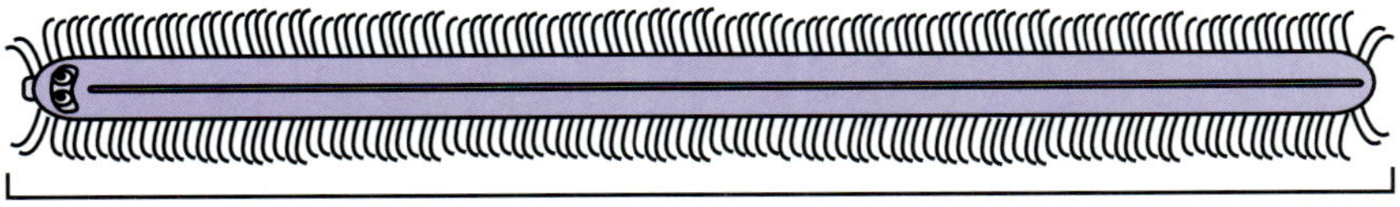
4. 
5. 
6. 
7. 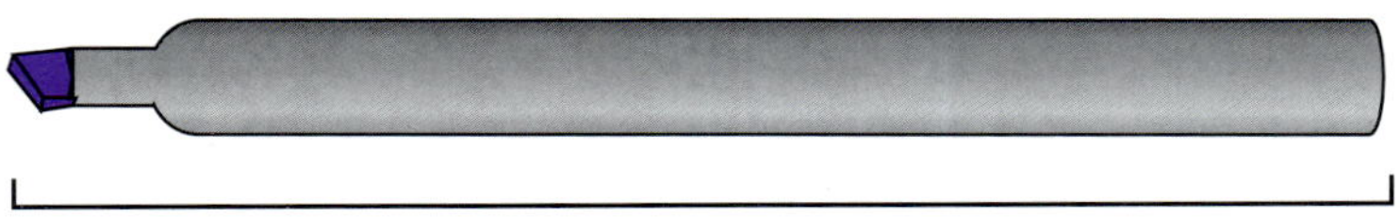
8. 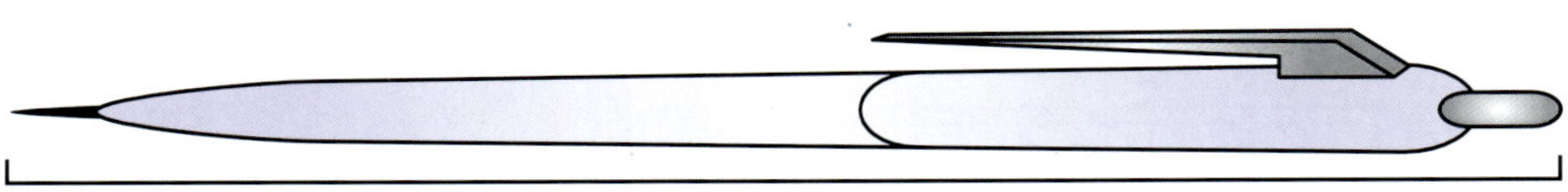
9. 
10. 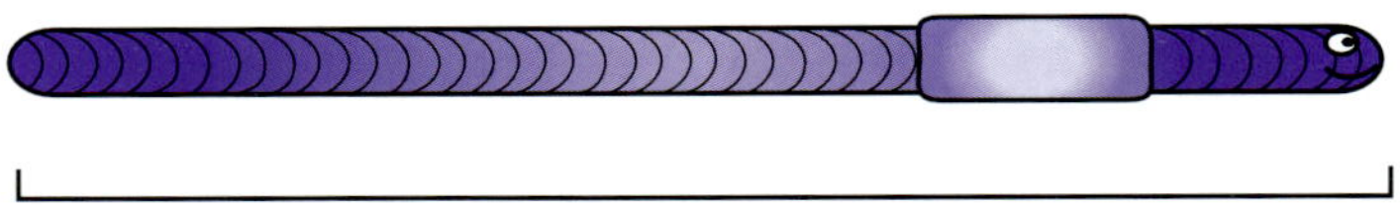
11. 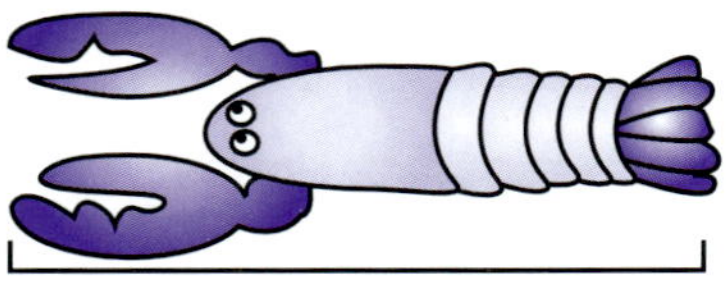
12. 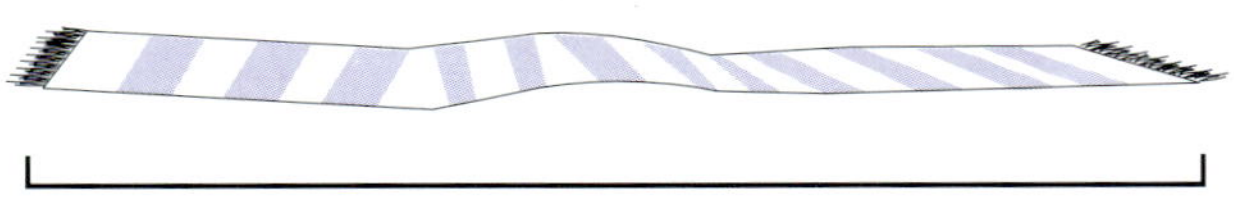

Answer Box

A	B	C	D	E	F
6 cm	3 cm	4 cm	2 cm	8 cm	7 cm
G	**H**	**I**	**J**	**K**	**L**
11 cm	5 cm	9 cm	10 cm	12 cm	13 cm

 Objective: Measure an object to the nearest centimeter.

Measure Up!

about 1 **centimeter (cm)**

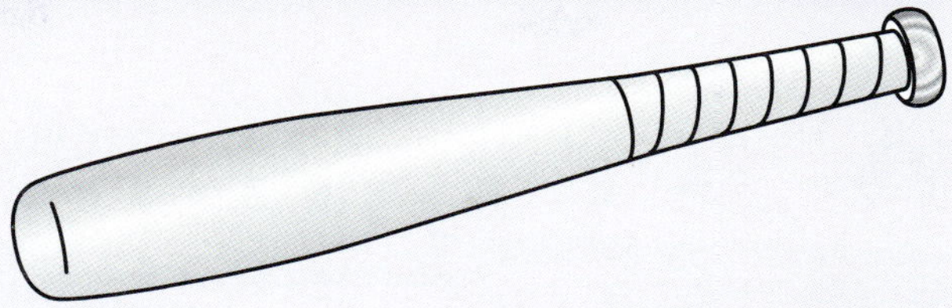

about 1 **meter (m)** long.

Choose the best estimate of the length.

1. your finger
   7 cm or 7 m
2. a crayon
   90 cm or 9 cm
3. an eraser
   5 cm or 5 m
4. a living room
   70 m or 7 m
5. a bucket
   5 m or 50 cm
6. a table
   90 cm or 9 cm
7. a small rug
   2 cm or 2 m
8. a playground
   7 cm or 70 m
9. a nickel
   2 cm or 2 m
10. a tree
    9 m or 9 cm
11. a house
    2 m or 20 m
12. a kitchen
    5 m or 5 cm

Answer Box

A	B	C	D	E	F
50 cm	20 m	5 m	70 m	2 m	9 m
G	**H**	**I**	**J**	**K**	**L**
5 cm	2 cm	90 cm	9 cm	7 m	7 cm

Objective: Name the best estimate for the length of an object.

Which Is Best?

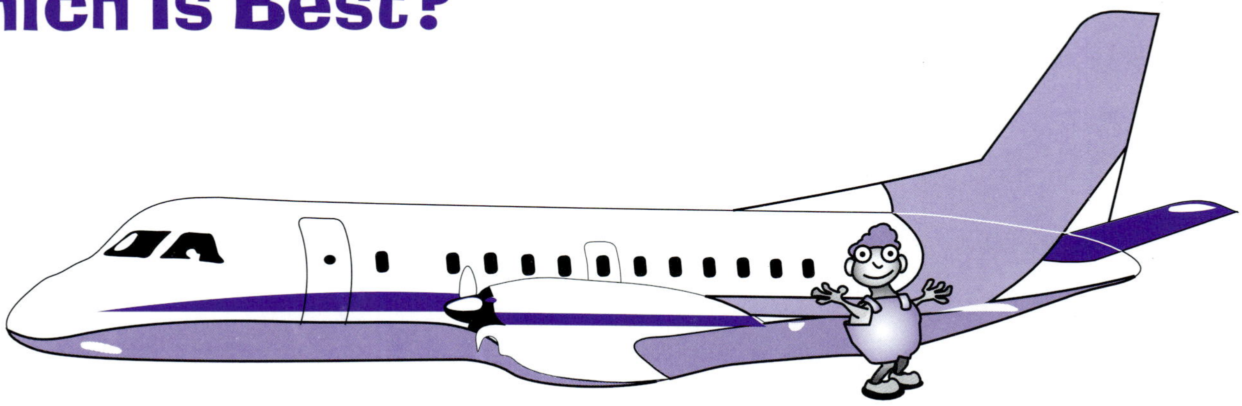

Choose the best estimate of the length.

1. a pen
15 cm or 15 m

2. a paperback book
18 cm or 80 cm

3. a desk
40 cm or 40 m

4. a fish hook
4 cm or 40 cm

5. a half dollar
30 cm or 3 cm

6. a nail
5 cm or 30 cm

7. a stool
18 m or 80 cm

8. a pond
40 m or 40 cm

9. a puppy
30 cm or 3 cm

10. an airplane
18 m or 80 cm

11. a swimming pool
15 cm or 15 m

12. a car
3 m or 30 cm

Answer Box

A	B	C	D	E	F
18 cm	3 m	30 cm	40 cm	15 cm	80 cm
G	**H**	**I**	**J**	**K**	**L**
40 m	5 cm	3 cm	15 m	18 m	4 cm

Objective: Name the best estimate for the length of an object.

All the Way Around!

Perimeter is the distance around a figure.

3 in. + 3 in. + 3 in. + 3 in. = 12 in.
The perimeter is 12 in.

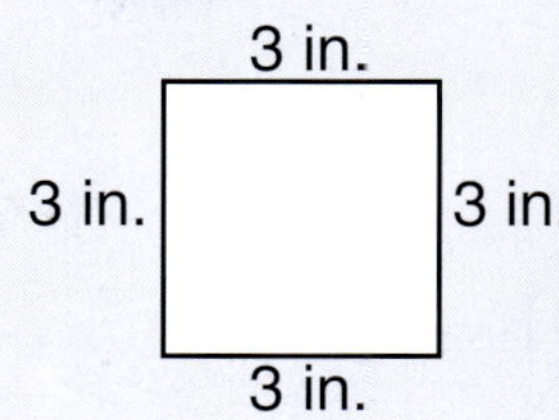

Find the perimeter.

1 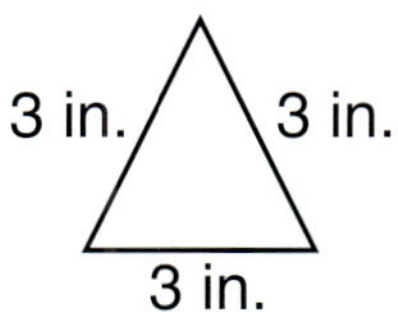

2 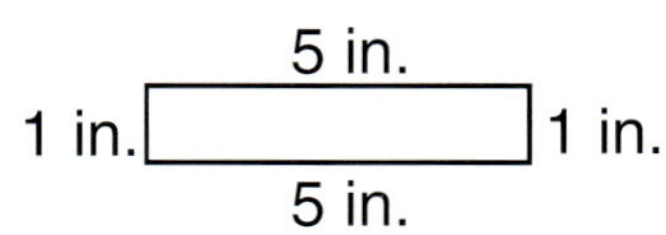

3 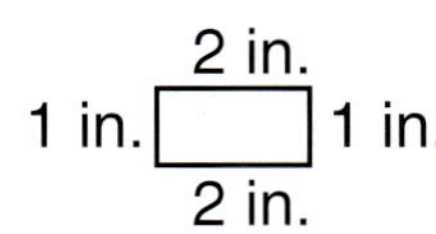

4 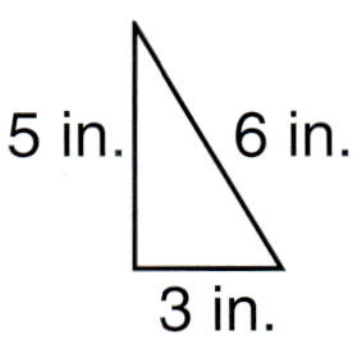

5 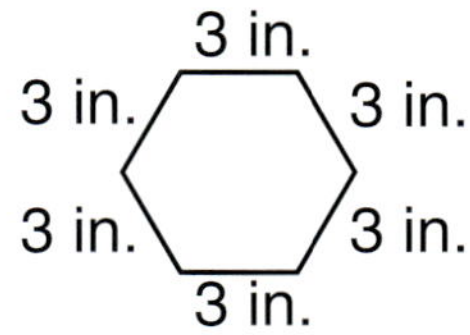

6 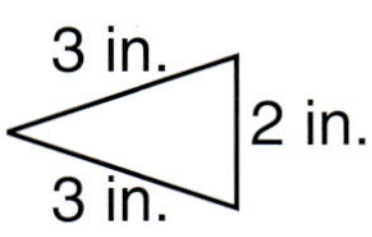

7 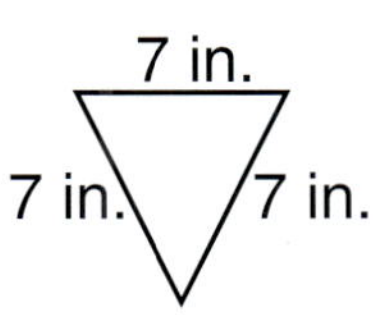

8 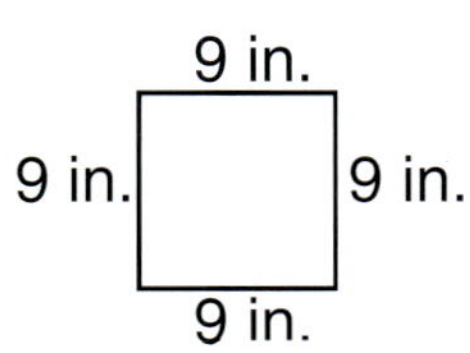

9 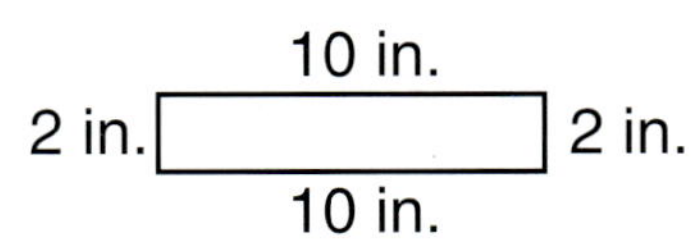

10 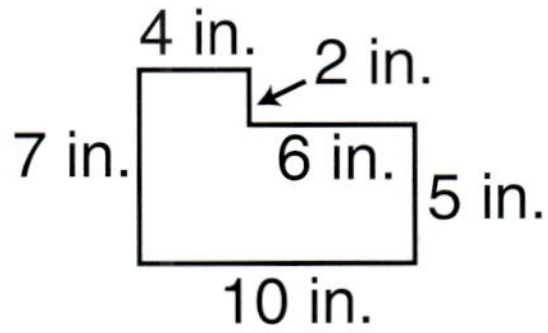

11 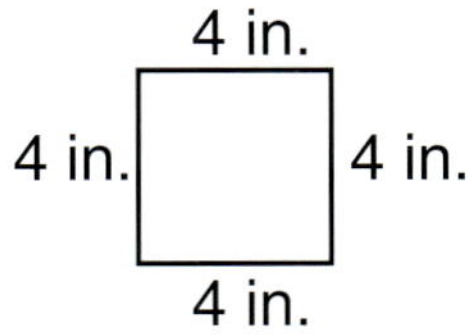

12 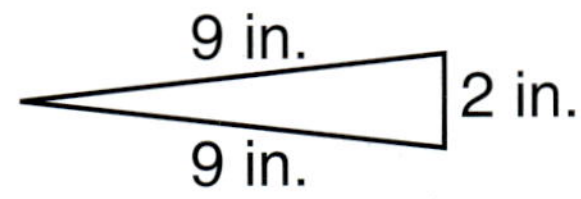

Answer Box

A	B	C	D	E	F
12 in.	8 in.	34 in.	21 in.	24 in.	6 in.
G	**H**	**I**	**J**	**K**	**L**
20 in.	36 in.	14 in.	9 in.	18 in.	16 in.

Objective: Find the perimeter of a given figure in inches.

Go the Distance!

Find the perimeter.

1. 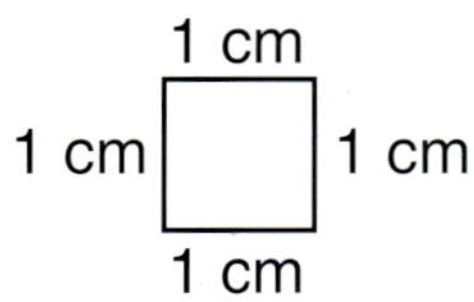

2. 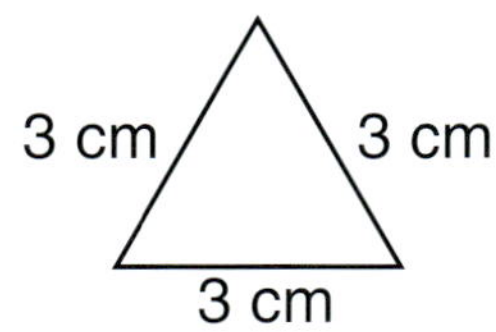

3. 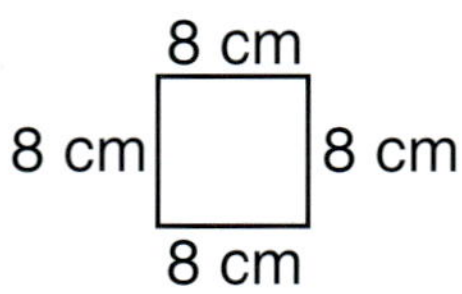

4. 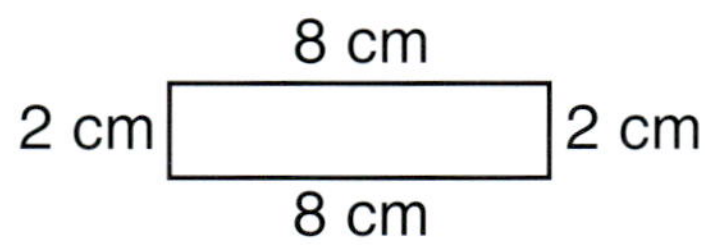

5. 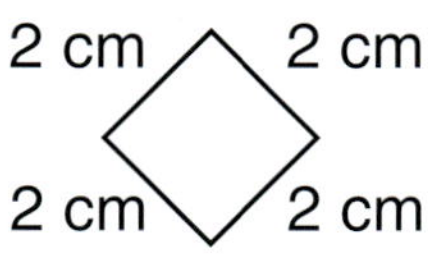

6. 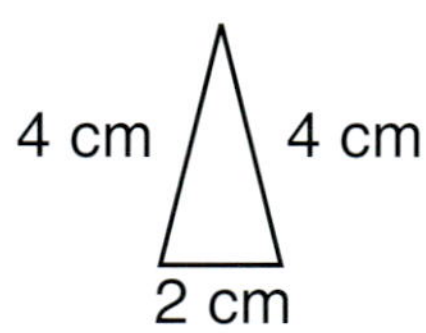

7. 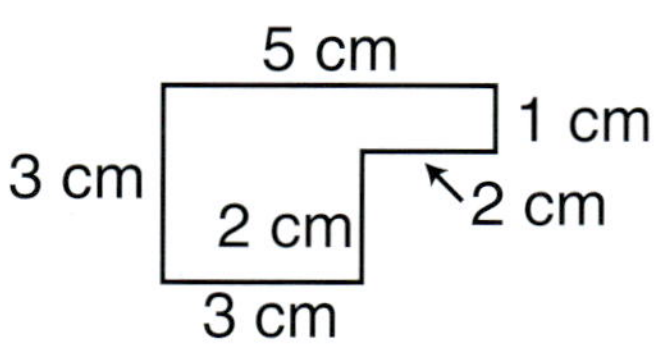

8. 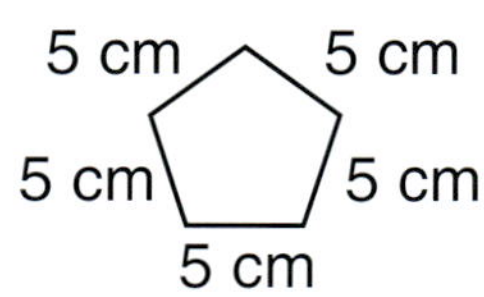

9. 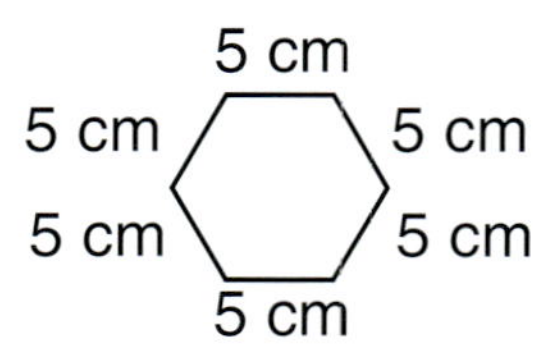

10. 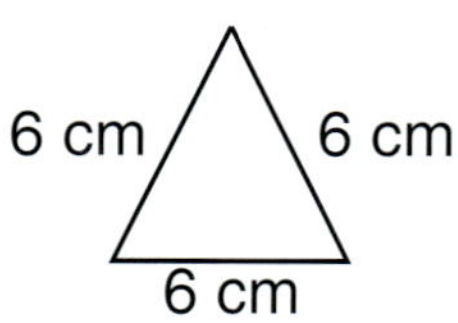

11. 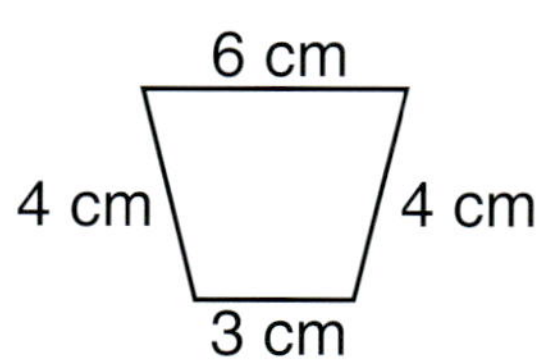

12. 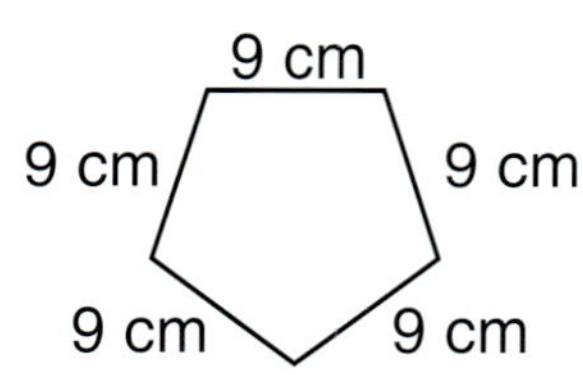

Answer Box

A	B	C	D	E	F
4 cm	20 cm	30 cm	32 cm	9 cm	10 cm
G	**H**	**I**	**J**	**K**	**L**
16 cm	18 cm	8 cm	17 cm	45 cm	25 cm

 Objective: Find the perimeter of a given figure in centimeters.

Getting Around

Find the perimeter.

1. 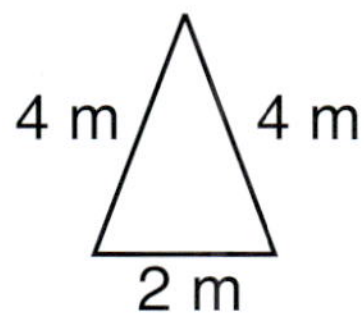

2. 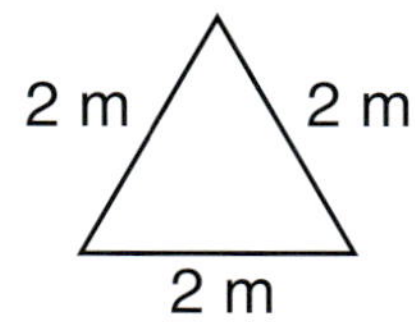

3. 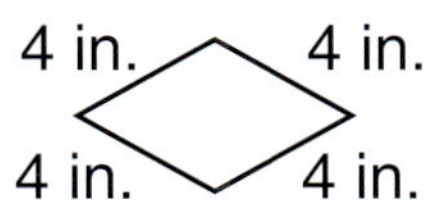

4. 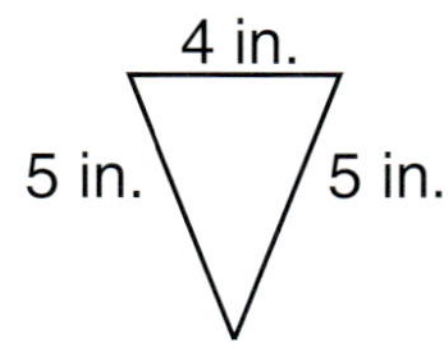

5. 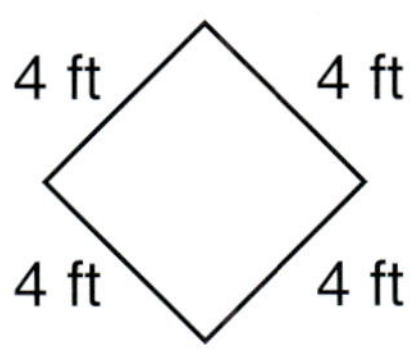

6. 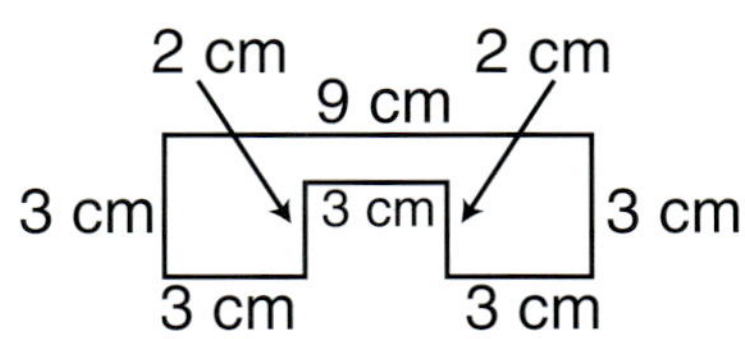

7. 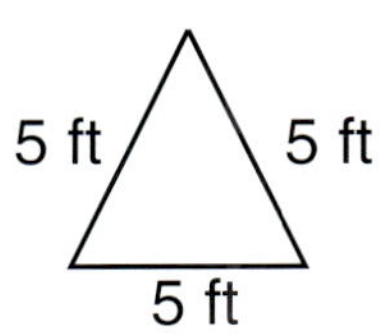

8. 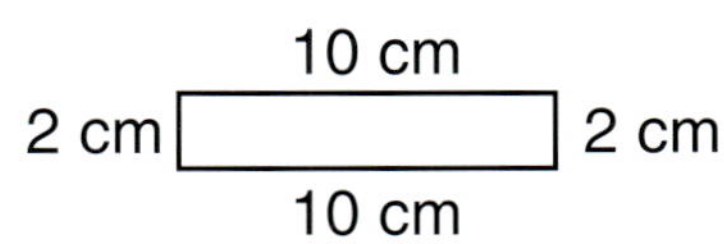

9. 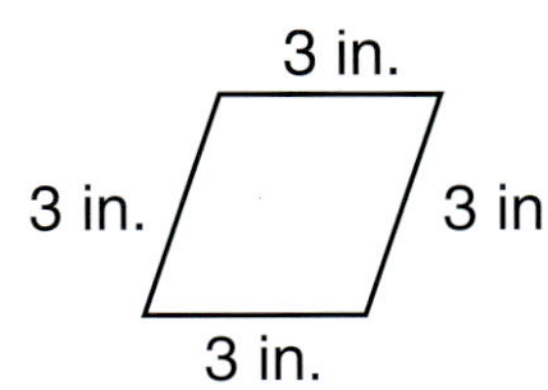

10. 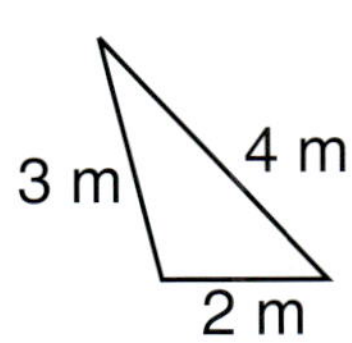

11. 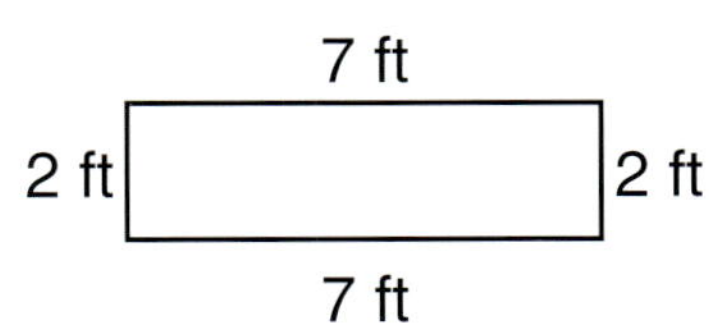

12. 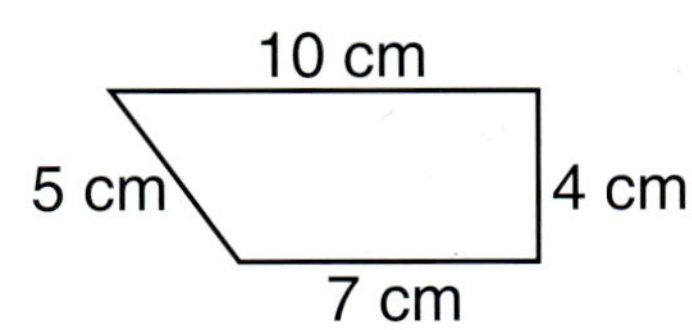

Answer Box

A	B	C	D	E	F
9 m	14 in.	6 m	26 cm	18 ft	16 in.
G	**H**	**I**	**J**	**K**	**L**
10 m	15 ft	24 cm	28 cm	12 in.	16 ft

Objective: Find the perimeter of a given figure.

What's Inside?

Area is the number of square units a figure covers.

The area is 6 square units.

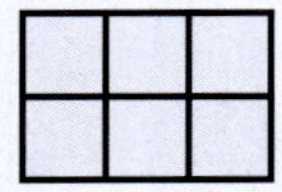

Find the area.

1 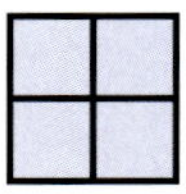

2 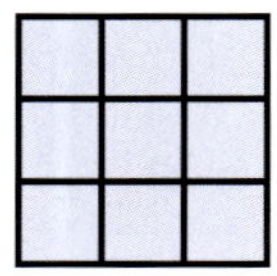

3 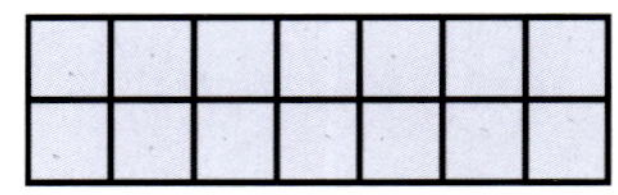

4 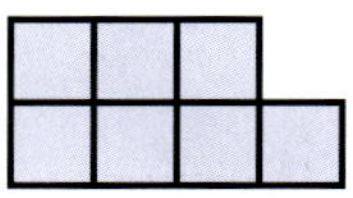

5 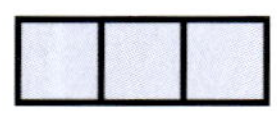

6 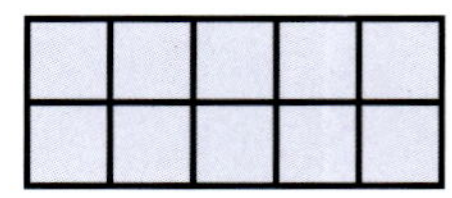

7 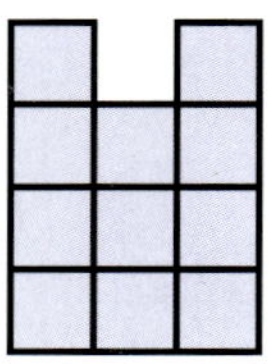

8 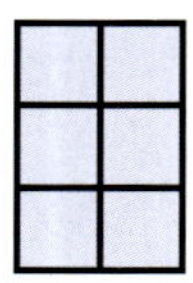

9 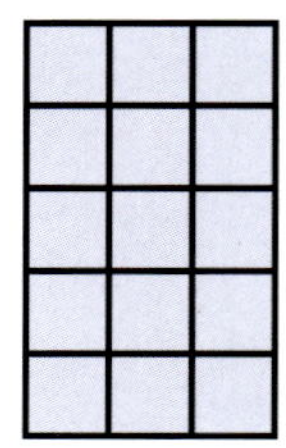

10 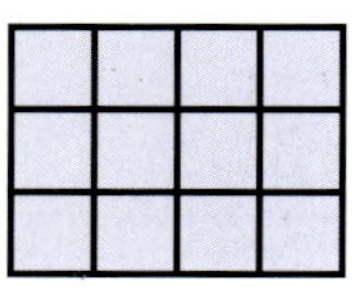

11 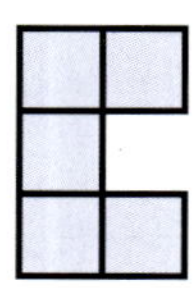

12 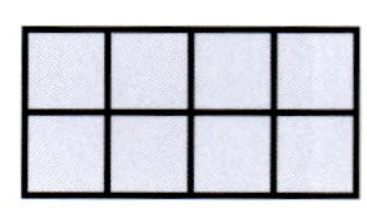

Answer Box

A	B	C	D	E	F
14 square units	5 square units	7 square units	12 square units	10 square units	6 square units
G	**H**	**I**	**J**	**K**	**L**
3 square units	15 square units	11 square units	4 square units	8 square units	9 square units

Objective: Count square units to find the area of a figure.

Counting Volume

Volume is the amount of space that an object takes up.

The volume is 8 cubic units.

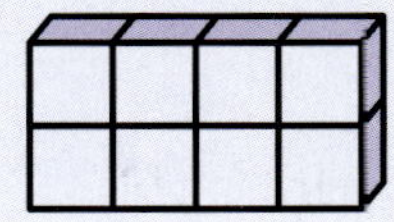

Find the volume.

1. 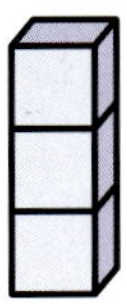

2. 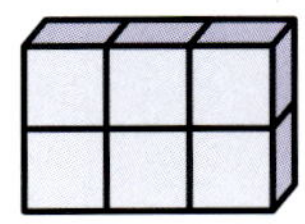

3. 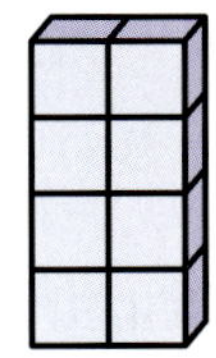

4. 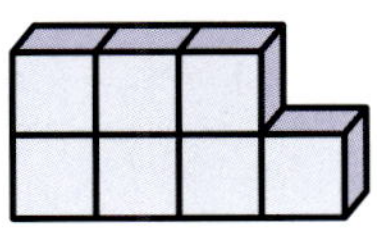

5. 

6. 

7. 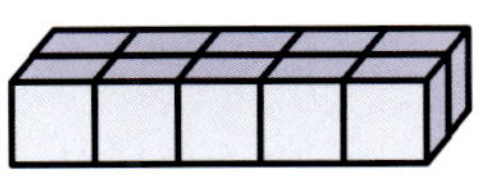

8. 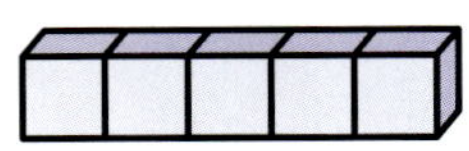

9. 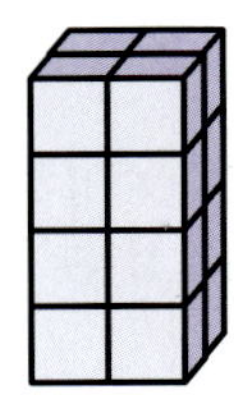

10. 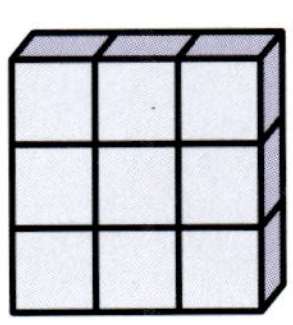

11. 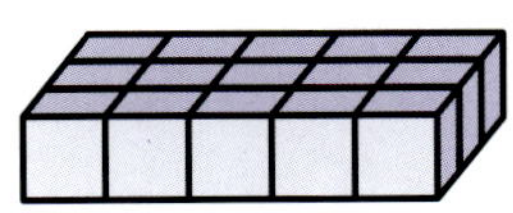

12. 

Answer Box

A	B	C	D	E	F
4 cubic units	15 cubic units	5 cubic units	18 cubic units	10 cubic units	9 cubic units
G	**H**	**I**	**J**	**K**	**L**
16 cubic units	8 cubic units	6 cubic units	11 cubic units	3 cubic units	7 cubic units

Objective: Count cubic units to find the volume of a figure.

Problem Solving: Using Logical Reasoning

Use the clues to solve the problem.

1. How many centimeters wide is the sandbox?

   The number is a 3-digit number with 6 tens.

   The sum of the digits is 24.

2. How many feet long is Kayla's classroom?

   The number is an even number less than 40.

   The sum of the two digits is 10.

3. How many points did the Rockets score during the game?

   The Rockets lost to the Tigers who scored 45 points.

   The ones digit is 4 more than the tens digit.

   The number is an odd number greater than 20.

4. What time does the bus for Newton leave?

   It leaves between 1:00 and 5:00.

   It leaves at a quarter to the hour.

   One of the minutes digits is the same as the hour digit.

5. How many inches long is Anita's aquarium?

   There are 4 tens in this 2-digit number.

   There are twice as many ones as tens.

6. How many cubic units is the locker?

   The number is greater than 100, but less than 200.

   The sum of the digits is 15.

   The tens digit is 7 more than the hundreds digit.

7 How many square units is the wading pool?

The number is between 50 and 100.

The sum of the digits is 12.

The difference of the digits is 4.

8 What is the perimeter of Nicole's insect collection box?

The perimeter is less than 100.

The digits are the same.

The sum of the digits is 2 more than 10.

9 How many pages are in a dictionary?

The number is between 400 and 800.

If 1 was added to the number, the hundreds digit would increase by 1.

The hundreds digit is a number you say when counting by 3.

10 How many shells did Jonathon collect?

If 10 was added to the number, the hundreds digit would increase by 1.

The sum of the hundreds and ones digits is 6.

The hundreds digit is 2 less than the ones digit.

11 What time does Lani get home from school?

She gets home between 3:00 and 4:00.

The sum of the minutes digits is 9.

One of the minutes digits is the same as the hour digit.

12 What time does Abdul go to baseball practice?

He goes after school.

The hour and the minutes digits are numbers you say when you count by 5.

The sum of the digits is 11.

Answer Box

A	B	C	D	E	F
5:15	66	48	699	84	294
G	**H**	**I**	**J**	**K**	**L**
28	186	37	3:36	969	4:45

Objective: Solve a problem by using logical reasoning.

Pour It Out!

1 cup (c)   1 pint (pt)   1 quart (qt)   1 gallon (gal)

Choose the best estimate.

1. a pitcher
   1 c or 1 qt
2. a coffee mug
   1 gal or 1 c
3. a watering can
   10 qt or 1 gal
4. a saucepan
   8 c or 8 qt
5. a birdbath
   2 c or 2 qt
6. a bathtub
   40 gal or 4 gal
7. a bucket
   8 qt or 80 gal
8. a juice bottle
   4 gal or 4 c
9. a vase
   2 pt or 20 c
10. a child's swimming pool
    8 qt or 80 gal
11. a tall glass
    2 c or 2 pt
12. a punch bowl
    4 gal or 4 c

Answer Box

A	B	C	D	E	F
40 gal	1 c	8 c	1 qt	1 gal	2 pt
G	**H**	**I**	**J**	**K**	**L**
4 gal	4 c	8 qt	80 gal	2 qt	2 c

 Objective: Select a capacity in cups, pints, quarts, or gallons for a named container.

That's About It!

Choose the best estimate.

1. juice glass
   1 c or 1 gal
2. small fishbowl
   6 pt or 6 gal
3. wading pool
   5 pt or 50 gal
4. large cooking pot
   50 gal or 5 gal
5. juice carton
   1 c or 1 qt
6. jug of juice
   1 gal or 1 qt
7. mixing bowl
   30 c or 3 qt
8. coffee pot
   5 pt or 5 gal
9. aquarium
   6 gal or 6 pt
10. wash basin
    6 gal or 6 qt
11. spaghetti sauce jar
    3 c or 30 c
12. bucket
    3 qt or 30 c

Answer Box

A	B	C	D	E	F
1 c	5 gal	3 c	1 qt	6 pt	1 gal
G	**H**	**I**	**J**	**K**	**L**
6 qt	3 qt	6 gal	50 gal	5 pt	30 c

Objective: Select a capacity in cups, pints, quarts, or gallons for a named container.

What's It Weigh?

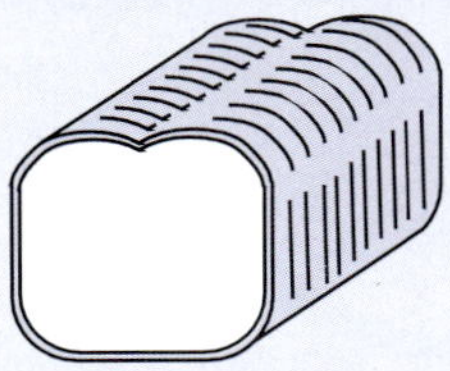

1 pound (lb)

1 ounce (oz)

Choose the best estimate.

1. a shoe
   10 lb or 1 lb
2. an orange
   6 oz or 6 lb
3. a soccer ball
   20 oz or 20 lb
4. a small dog
   10 lb or 1 lb
5. a television set
   6 lb or 60 lb
6. a pen
   2 oz or 20 oz
7. a watermelon
   6 oz or 6 lb
8. a crayon
   1 lb or 1 oz
9. a small bunch of grapes
   4 oz or 40 lb
10. a chair
    20 oz or 20 lb
11. a hardcover book
    40 oz or 40 lb
12. a medium-size dog
    40 lb or 40 oz

Answer Box

A	B	C	D	E	F
20 oz	4 oz	20 lb	10 lb	2 oz	6 oz
G	**H**	**I**	**J**	**K**	**L**
60 lb	40 oz	1 lb	6 lb	40 lb	1 oz

 Objective: Select a weight in ounces or pounds for a named object.

Weigh It!

Choose the best estimate.

1. pair of boots
   5 lb or 5 oz
2. bag of potatoes
   10 oz or 10 lb
3. kitten
   3 oz or 3 lb
4. baby
   8 lb or 80 lb
5. eraser
   10 lb or 1 oz
6. hamburger
   8 oz or 8 lb
7. box of cereal
   5 lb or 15 oz
8. table
   3 lb or 30 lb
9. mitten
   3 oz or 30 lb
10. child's scissors
    5 oz or 5 lb
11. paperback book
    10 oz or 1 oz
12. large dog
    8 oz or 80 lb

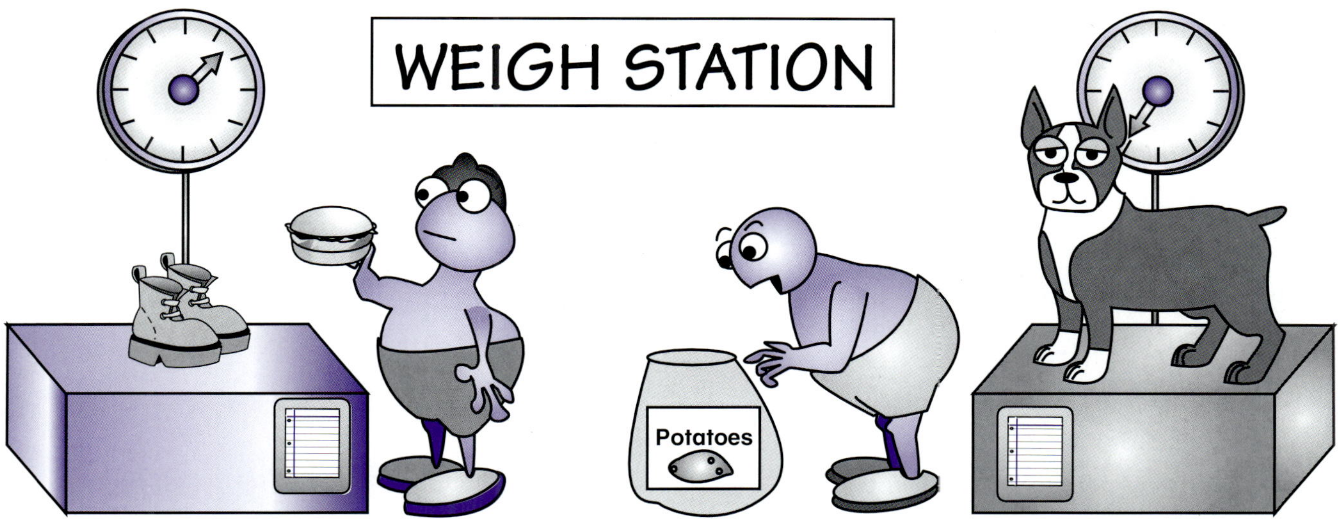

Answer Box

A	B	C	D	E	F
3 lb	3 oz	80 lb	10 lb	8 lb	5 lb
G	**H**	**I**	**J**	**K**	**L**
10 oz	1 oz	8 oz	30 lb	5 oz	15 oz

Objective: Select a weight in ounces or pounds for a named object.

Hold It!

1 milliliter (mL)

1 liter (L)

Choose the best estimate.

1. a dog's water bowl
   1 L or 10 mL

2. a teacup
   250 mL or 250 L

3. an ice cream carton
   20 mL or 2 L

4. a cooking pot
   5 L or 5 mL

5. a wading pool
   250 mL or 250 L

6. a small spoon
   5 L or 5 mL

7. a washing machine
   250 mL or 25 L

8. a soup spoon
   20 mL or 2 L

9. a bucket
   20 mL or 20 L

10. a raindrop
    1 L or 1 mL

11. soup ladle
    1 L or 100 mL

12. a kitchen sink
    5 L or 50 L

Answer Box

A	B	C	D	E	F
20 mL	50 L	1 L	5 L	100 mL	250 L
G	**H**	**I**	**J**	**K**	**L**
5 mL	250 mL	1 mL	25 L	2 L	20 L

Objective: Select a capacity in milliliters or liters for a named container.

Lots of Liters

Choose the best estimate.

1. punch bowl
   10 L or 10 mL
2. milk glass
   250 mL or 25 mL
3. pail
   15 L or 150 L
4. gas tank of small car
   3 L or 30 L
5. large pitcher
   3 L or 30 mL
6. juice glass
   15 L or 150 mL
7. ice cube
   30 mL or 30 L
8. eyedropper
   1 L or 10 mL
9. teaspoon
   25 L or 25 mL
10. tea kettle
    1 L or 10 L
11. bathroom sink
    25 L or 250 mL
12. bathtub
    15 L or 150 L

Answer Box

A	B	C	D	E	F
1 L	30 mL	25 L	3 L	10 mL	150 L
G	**H**	**I**	**J**	**K**	**L**
30 L	10 L	15 L	25 mL	150 mL	250 mL

Objective: Select a capacity in milliliters or liters for a named container.

Mass Measures

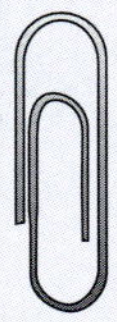

1 gram (g)

1 kilogram (kg)

Choose the best estimate.

1. sofa
   1 kg or 100 kg
2. a quarter
   5 g or 5 kg
3. a small dog
   5 kg or 50 g
4. a bee
   2 g or 2 kg
5. a boot
   10 kg or 1 kg
6. a hardcover book
   2 kg or 20 kg
7. a small television
   10 kg or 100 kg
8. a bicycle
   150 g or 15 kg
9. a mug
   150 g or 15 g
10. a necklace
    5 g or 50 g
11. a computer
    2 kg or 20 kg
12. a slice of bread
    15 g or 150 g

Answer Box

A	B	C	D	E	F
5 kg	150 g	2 g	100 kg	2 kg	5 g
G	**H**	**I**	**J**	**K**	**L**
20 kg	1 kg	50 g	10 kg	15 kg	15 g

 Objective: Select a mass in grams or kilograms for a named object.

Mass Matters

Choose the best estimate.

1. safety pin
   1 g or 1 kg
2. bed
   7 kg or 70 kg
3. bag of oranges
   30 kg or 3 kg
4. truck
   2,000 kg or 2,000 g
5. pen
   30 g or 3 kg
6. balloon
   1 g or 10 g
7. can of coffee
   10 g or 1 kg
8. cat
   7 kg or 70 g
9. paperback book
   2,000 kg or 200 g
10. worm
   3 g or 3 kg
11. puppy
   200 g or 2,000 g
12. table
   30 kg or 30 g

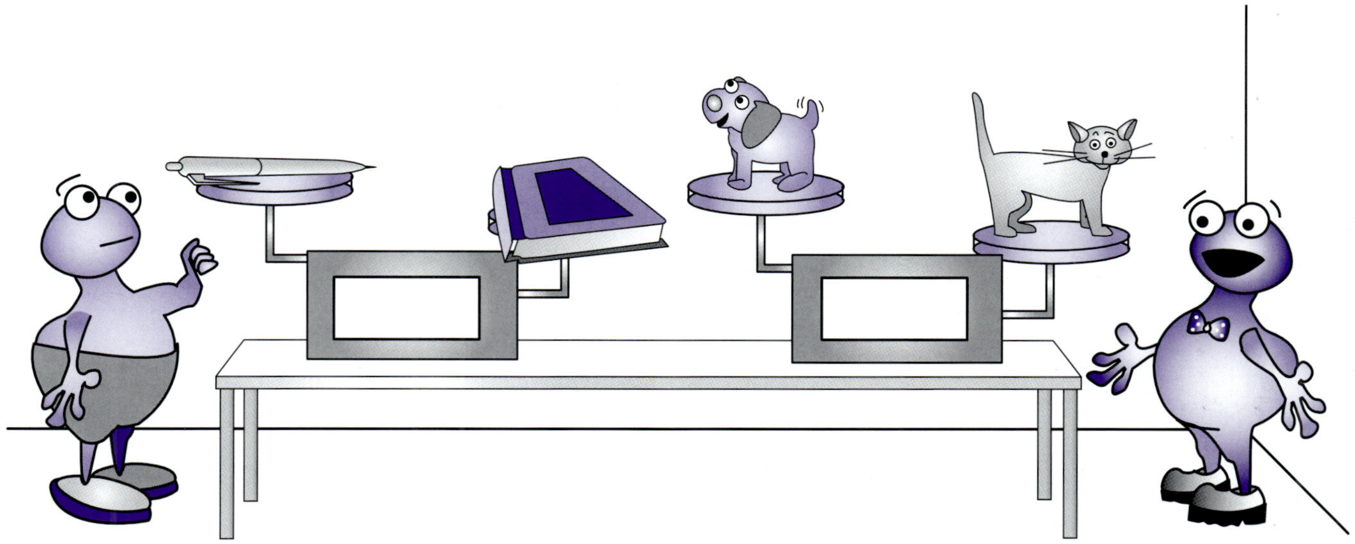

Answer Box

A	B	C	D	E	F
200 g	3 kg	7 kg	30 kg	2,000 kg	1 g
G	**H**	**I**	**J**	**K**	**L**
30 g	2,000 g	70 kg	10 g	3 g	1 kg

Objective: Select a mass in grams or kilograms for a named object.

Name That Figure!

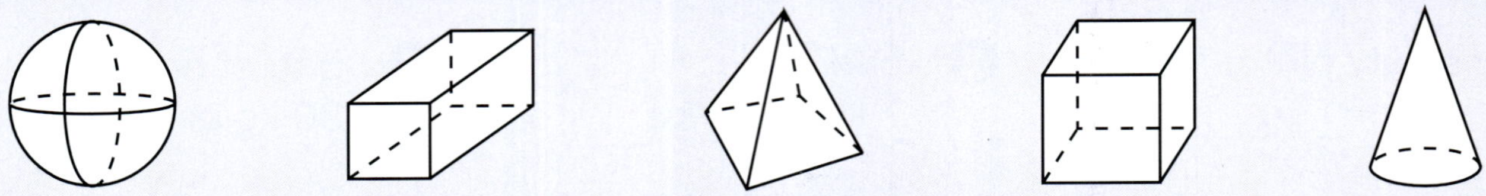

Find the figure name that makes the statement true.

1. A 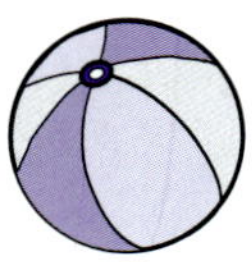 is a ■.

2. A  is a ■.

3. A  is a ■.

4. A  is a ■.

5. A  is a ■.

6. A  is a ■.

Find the object that makes the statement true.

7. An ■ is a cone.
8. A ■ is a rectangular prism.
9. A ■ is a sphere.
10. A ■ is a pyramid.
11. A ■ is a cylinder.
12. A ■ is a cube.

Answer Box

A	B	C	D	E	F
Cylinder	Cone	Sphere		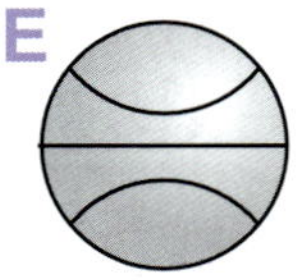	Rectangular prism
G	**H**	**I**	**J**	**K**	**L**
			Pyramid	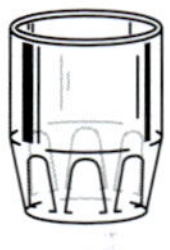	Cube

 Objective: Identify the 3-dimensional figure that matches a given pictured object.

3-D Descriptions

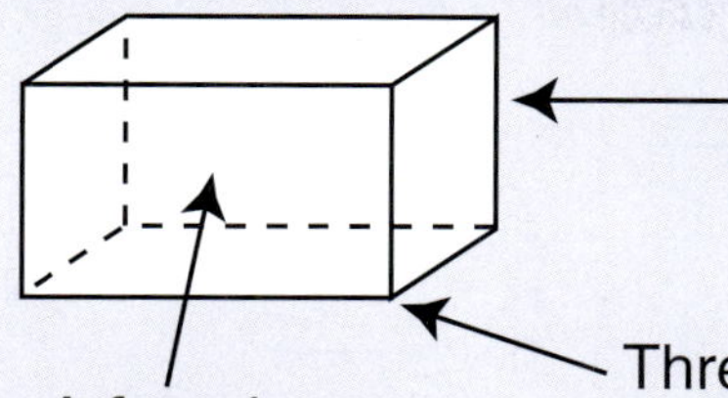

Find the number or word that makes the statement true.

1. A rectangular prism has ■ faces.
2. A cube has ■ edges.
3. A cone has 1 ■ face.
4. A sphere has ■ faces.
5. A square pyramid has ■ corners.
6. A square pyramid has ■ edges.
7. A cube has 6 ■ faces.
8. A cylinder has ■ faces.
9. A ■ has 12 edges.
10. A cylinder has 0 corners and 0 ■.
11. A rectangular prism has 8 ■.
12. A ■ has 5 faces.

Answer Box

A	B	C	D	E	F
0	6	Flat	Corners	8	12
G	**H**	**I**	**J**	**K**	**L**
Edges	Square	5	Rectangular prism	2	Pyramid

Objective: Describe a 3-dimensional figure in terms of its faces, edges, and corners.

Figure It Out!

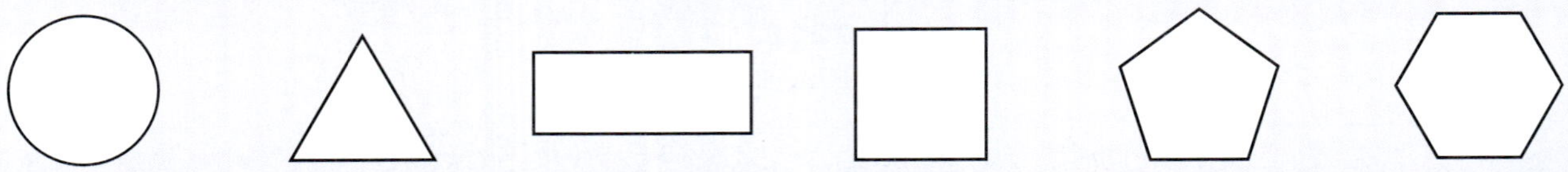

circle triangle rectangle square pentagon hexagon

Find the figure name that makes the statement true.

1. A  is a ■.
2. A  is a ■.
3. A  is a ■.
4. A  is a ■.
5. A  is a ■.
6. A 

is a ■.

Find the shape that makes the statement true.

7. A ■ is a square.
8. A ■ is a rectangle.
9. A ■ is a circle.
10. A ■ is a pentagon.
11. A ■ is a hexagon.
12. A ■ is a triangle.

Answer Box

A	B	C	D	E	F
Circle	Hexagon	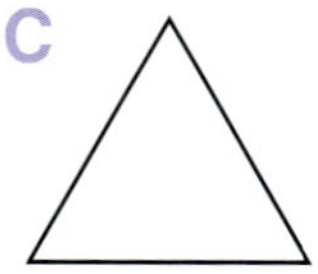	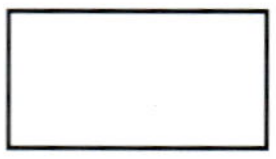	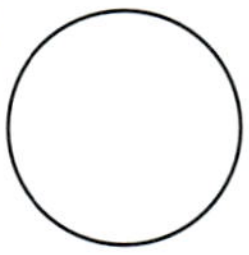	Triangle
G	**H**	**I**	**J**	**K**	**L**
	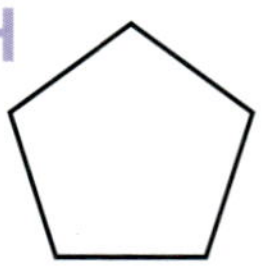	Pentagon	Rectangle	Square	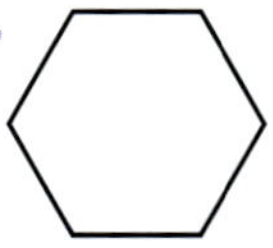

 Objective: Identify the 2-dimensional figure.

Down to Details!

Find the number or word that makes the statement true.

1. A rectangle has ■ corners.
2. A circle has ■ sides.
3. A hexagon has ■ sides.
4. A triangle has ■ corners.
5. A pentagon has ■ corners.
6. A square has 4 sides of the ■ length.
7. A rectangle has 2 short sides of the same length and ■ long sides of the same length.
8. A triangle has 3 sides of the same length or 3 sides of ■ lengths.
9. A ■ has 0 corners.
10. A ■ has 5 sides.
11. A ■ has 6 corners.
12. A rectangle and a ■ have the same number of corners.

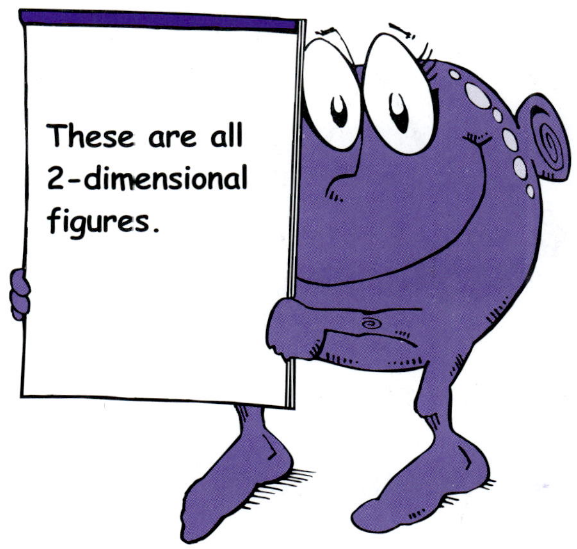

Answer Box

A	B	C	D	E	F
6	Circle	2	Pentagon	Same	0

G	H	I	J	K	L
5	Hexagon	4	3	Square	Different

Objective: Describe a 2-dimensional figure in terms of the number of sides, lengths of sides, and number of corners.

Problem Solving: Finding a Pattern

Find the answer.

Ted is planting a garden. He plants 5 flowers in the first row, 7 flowers in the second row, and 9 flowers in the third row.

1. If Ted continues this pattern, how many flowers will he plant in the fourth row?
2. If Ted continues this pattern, how many flowers will he plant in the sixth row?

Latoya is knitting a striped scarf. To make the first four stripes, she knits 7 red rows, 3 blue rows, 8 red rows, and 4 blue rows.

3. If Latoya continues this pattern, how many rows will she knit to make the fifth stripe?
4. If Latoya continues this pattern, how many rows will she knit to make the eighth stripe?

Nolan is setting up chairs for a concert. He puts 4 chairs in the first row, 3 chairs in the second row, 6 chairs in the third row, and 5 chairs in the fourth row.

5. If Nolan continues this pattern, how many chairs will he put in the sixth row?
6. If Nolan continues this pattern, how many chairs will he put in the eleventh row?

Paco is pouring drinks at a fair. He pours 2 lemonades, 1 orange drink, 3 lemonades, and 1 orange drink.

7. If Paco continues this pattern, how many lemonades will he pour next?

8. If Paco continues this pattern, how many orange drinks will he pour next?

Ms. Turner is making a quilt for her daughter. She puts a star in squares 2, 4, and 6. She puts a heart in squares 3, 6, and 9.

9. If Ms. Turner continues this pattern, in what square will she put the next star?

10. There is a star and a heart in square 6. If Ms. Turner continues this pattern, what is the next square in which she will put a star and a heart?

Ruth is selling used books and toys at a garage sale. During the first hour she made $30. During the second hour she made $28. During the third hour she made $25.

11. If this pattern of sales continues, how many dollars will Ruth make during the fifth hour?

12. If this pattern of sales continues, how many dollars will Ruth make during the seventh hour?

Answer Box

A	B	C	D	E	F
16	7	1	3	12	4
G	**H**	**I**	**J**	**K**	**L**
8	9	15	14	11	6

Objective: Use a number pattern to solve a problem.

Lines and More

line | line segment | ray | angle

Find the number or word that makes the statement true.

1. A is a ■.
2. A is a ■.
3. A 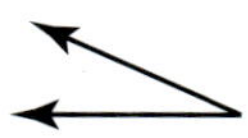 is an ■.
4. A is a ■.
5. A  has ■ line segments.
6. A 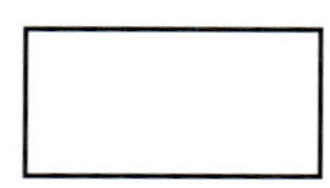 has ■ angles.
7. A 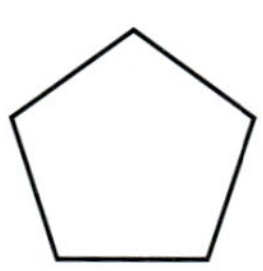 has ■ line segments.
8. A 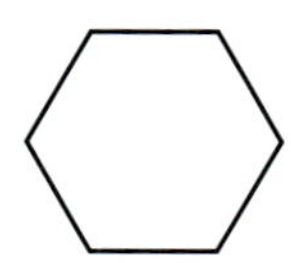 has ■ angles.

Find the figure that makes the statement true.

9. ■ is a ray.
10. ■ is an angle.
11. ■ is a line.
12. ■ is a line segment.

Answer Box

A	B	C	D	E	F
6			Angle	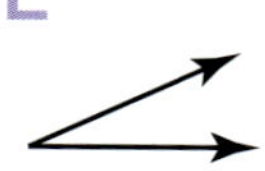	5
G	**H**	**I**	**J**	**K**	**L**
4	Ray	3		Line	Line segment

 Objective: Identify lines, line segments, rays, and angles.

Just a Part of Me

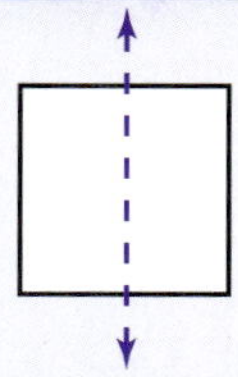

line of symmetry

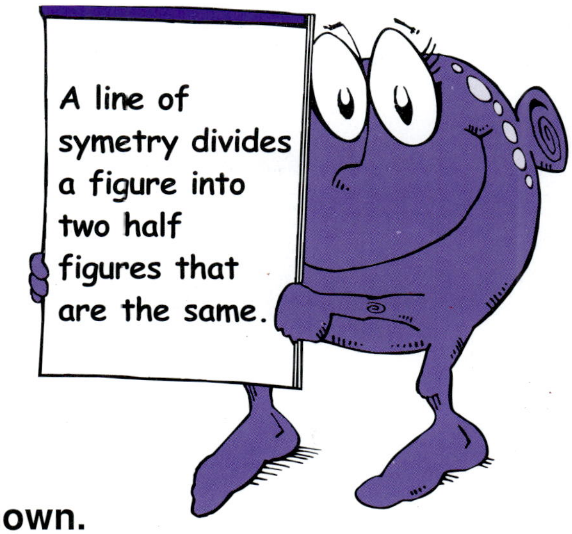

The line of symmetry and half of the figure is shown. Find the complete figure.

1 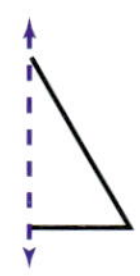

2 

3 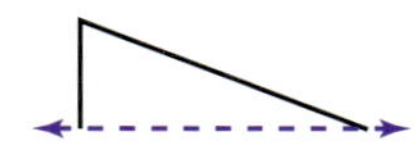

4 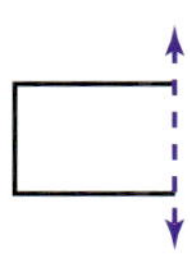

5 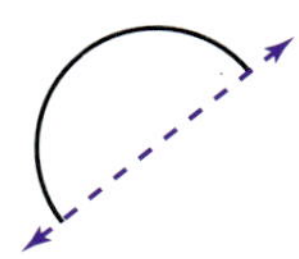

6 

7 

8 

9 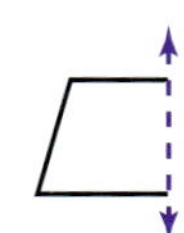

10 

11 

12 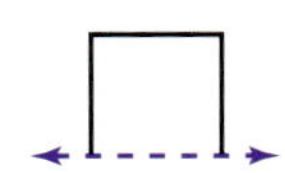

Answer Box

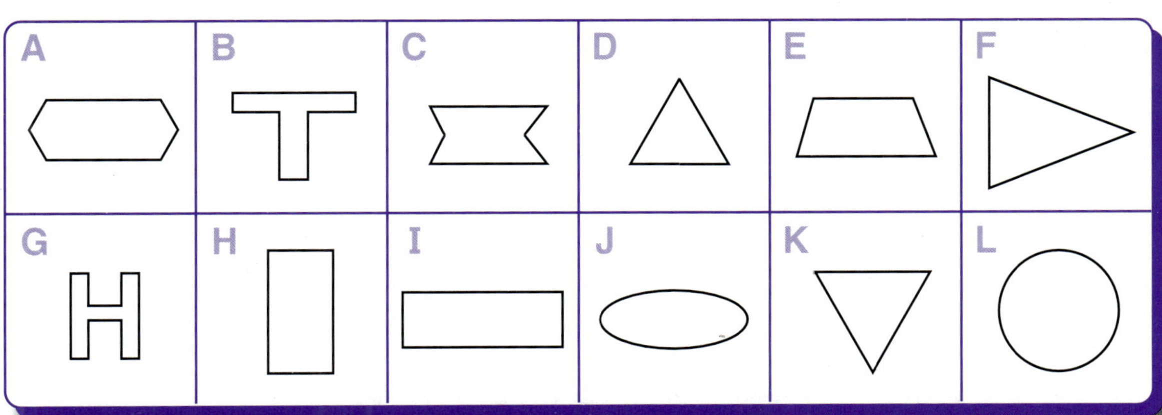

Objective: Match a partial figure to a complete figure that shows a line of symmetry.

Many Lines!

This figure has 2 lines of symmetry.

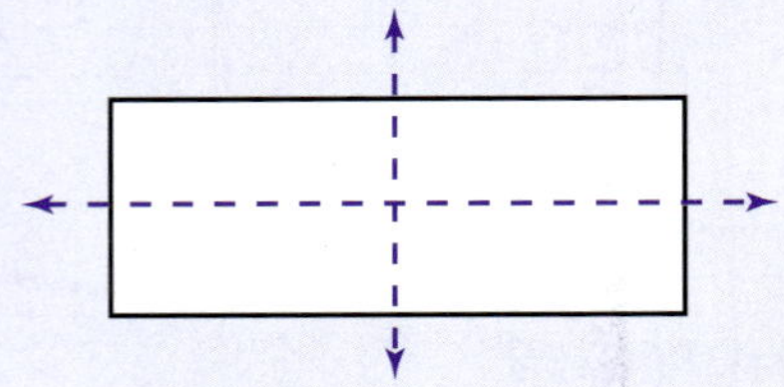

How many lines of symmetry does the figure have?

1. 
2. 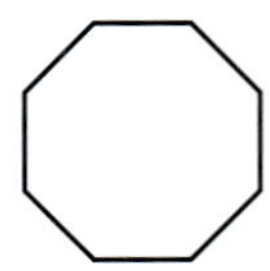
3. 
4. 
5. 
6. 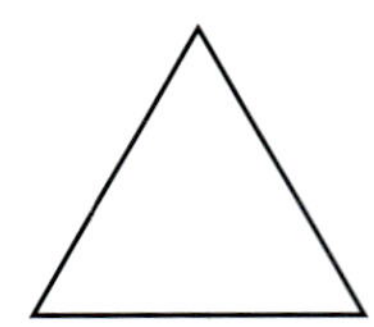
7. 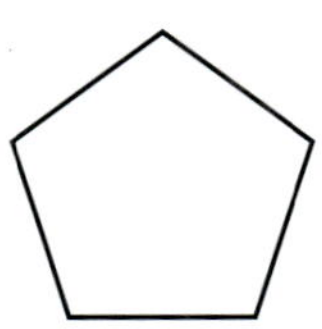
8. 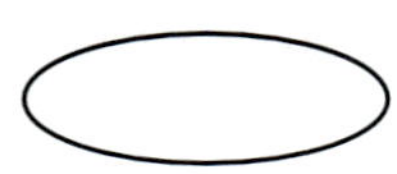
9. 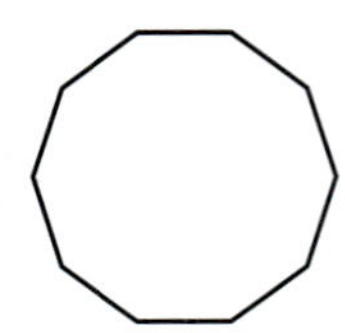
10. 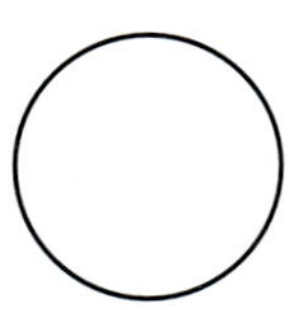
11. 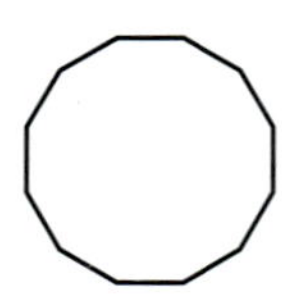
12. 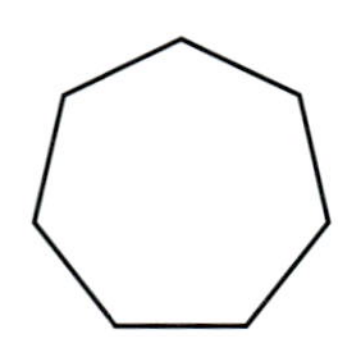

Answer Box

A	B	C	D	E	F
Too many to count	6	1	10	8	7
G	**H**	**I**	**J**	**K**	**L**
4	5	12	0	2	3

 Objective: Tell how many lines of symmetry a figure has.